Women in Water

Women in Water

How to Support Everyone's Contribution

Cheryl Davis, Andrée De Cock,
Arlinda Ibrahimllari, Liudmyla Odud
and Siyka Radilova

IWA PUBLISHING

Published by **IWA Publishing**
Unit 104–105, Export Building
1 Clove Crescent
London E14 2BA, UK
Telephone: +44 (0)20 7654 5500
Fax: +44 (0)20 7654 5555
Email: publications@iwap.co.uk
Web: www.iwapublishing.com

First published 2024

British Library Cataloguing in Publication Data
A CIP catalogue record for this book is available from the British Library

ISBN: 9781789064063 (paperback)
ISBN: 9781789064070 (eBook)
ISBN: 9781789064087 (ePub)

Doi: 10.2166/9781789064070

This eBook was made Open Access in March 2024.

Endorsements for *Women in Water*

'What an amazing read I had, marvelling through the excerpts of speeches by amazing women around the world! I feel as if I am floating from country to country, mulling through opinions coming from each and everyone's mind and heart touching upon various issues concerning women and young professionals within the water industry.

Huge applause to the entire team for this and I can't wait for it to be published, for me to share with my men and women colleagues alike.'

Norhayati Abdullah PhD
Associate Professor/Associate Director
UTM Universiti Teknologi Malaysia

'As a father of three daughters, this book resonates with me, emphasizing the urgent need to amplify women's roles in the water sector. More than just an equity issue, leveraging the talents of women is essential in addressing the pressing challenges of water supply, quality, and climate change. Reflecting men's current dominance in industries, including water, as well as the cultural patterns that reinforce this dominance, the voices captured in the book call for substantial change through a collective shift in behavior. Beyond this imperative, the book movingly portrays women's involvement as crucial and inspiring. The book clearly identifies the need for men and women in the water industry to actively support women's inclusion and empowerment, fostering transformative change in the water sector.'

Carlos A. Bustos, M.S.
Water Conservation Manager
Albuquerque Bernalillo County Water Utility Authority, USA
Member of the IWA Sustainability Committee in the Water Sector

'Thank you for allowing me the opportunity to edit this wonderful and relevant book.

This book has done so much for me as an Individual whilst reading. I learned so much and was able to relate to my female colleagues as they told their stories. I am so excited about this publication!'

Sharon Archie
Manager, Water Resources, Water Resources Agency
Water and Sewerage Authority, Trinidad and Tobago
Member of the IWA Sustainability Committee in the Water Sector

Acknowledgments

Many individuals and organizations have contributed their time, energy and resources to make this book possible. Speakers shared their experiences, challenges and lessons learned with intelligence, generosity and courage. Valuable research findings came from individuals who completed surveys and participated in webinars. Webinars and a workshop were supported by moderators, International Water Association members and International Water Association staff.

Special thanks go to the African Water and Sanitation Association (AFWASA) which supported recruitment for the webinar composed of women in the water industry in Africa, and to Agua y Saneamientos Argentinos (AYSA), which provided a speaker for a webinar, a moderator and speaker for a workshop, and numerous excellent photographs illustrating the valuable contributions of women in the water industry. IWA Publishing has worked with us to make the book readable.

It is a privilege to be useful, and the authors of this book appreciate everyone who has helped us convert our intentions into a product that we hope will be useful to others.

Cheryl Davis
Chair of the International Water Association's Specialist Group on Sustainability in the Water Sector

Contents

Foreword by Tom Mollenkopf

When I learned of the proposed publication of this book, I asked the editors if I could assist in some way. I did this because when it comes to gender equity, the water sector – like so many other sectors around the world – has a problem.

Put simply, women are under-represented in our sector, especially in leadership positions. I am convinced that we in water are committed to the concept of people being employed based on merit, without regard to gender. But there are many barriers that contribute to an ongoing gender imbalance. These may be subtle constraints like lack of flexible and inclusive employment experiences, or unwelcoming recruitment and promotion practices. But sometimes it is as simple as gender discrimination – an entrenched view that women are inherently unsuited for certain roles.

Whatever the cause, gender imbalance impoverishes us all, and that makes this publication vitally important. First, by facilitating the full engagement of women in our workplaces (and society more broadly), we ensure that we have the benefit of 50% of the world's talent, intellect and insight; talent that may otherwise be lost to us or underutilised by being excluded from the workforce. Second, an inclusive and diverse workforce brings different ways of thinking and fresh perspectives – it increases creativity and leads to better decision making. And finally, gender balance is fair: it leads to a more just and inclusive society, where everyone has an opportunity to succeed.

When I offered to contribute to this publication, I was also motivated by the belief that working for gender equity should not be the responsibility of women alone. It is essential that men take a role in being champions of change, adding their voice, and taking action through their own behaviours and organisational practices.

I commend the editors and contributors to this project and look forward to continuing to work towards a fairer and more inclusive water sector.

Tom Mollenkopf AO
President, International Water Association
Australia

Foreword by Diane D'Arras

I am very pleased to write an introduction for a book that addresses a topic that is critical in terms of our ability to adequately address the daunting challenges faced by the water industry. Both my work in the private sector and my experience as a Past President of the International Water Association have made me acutely aware that the talents and abilities of women are not always fully utilized. As an industry, we have the opportunity and responsibility to acknowledge the consequences of historical attitudes and practices, and promote a collaborative culture that builds, uses, and appreciates everyone's talents.

I would like to thank all of the women who have spoken to us through this book. Changing mentalities will still be a long, long, road, but we start down the path by acknowledging the world as it is today. It takes several generations in social rights between the realization that change is needed, and the achievement of a new reality. I am confident that women are mobilizing now, and that men will join us in this useful effort.

Diane D'Arras
Past President of the International Water Association (IWA)
France

Foreword by Sylvain Usher

Throughout history, women have been highly concerned about meeting their family's water needs. Comparatively, men have attained mastery at the technical level, while women have consistently ensured the proper use of this precious liquid (water) at the household level in ever-increasing populations. Concerning contemporary advancement, women equally venture into the technological acumen of water harvesting, treatment, supplies, as well as sanitation. However, resistance to change, rooted in cultural practices and attitudes, abounds especially regarding water sanitation consciousness. In Africa, gender balance in the Water, Sanitation and Hygiene (WASH) sector is a challenge for which we, the African Water and Sanitation Association, have taken radical measures to advocate for change in mentalities and practices. Although women do not hold positions in the management of the WASH sector, the exceptional management qualities, and resilience of the few women in the sector during crises have been remarkable.

This publication of *Women in Water* at such a time is appropriate, because it records the shared experiences of women working in the sector globally. It is my singular honor to appreciate all these remarkable women who, out of their own volition, expressed interest in sharing their professional experiences in the water and sanitation industry, culminating in the compilation of this book. This book, therefore, does not only inspire women who want to embrace the sector to further their professionalism, but also debunks myths about women's ability to assume highly contested responsible positions in the water and sanitation sector at local, regional, and global levels. Regrettably, due to gender clashes, the road to achieving the feat envisioned for women may certainly be long, yet a journey of a thousand miles begins with one step. I therefore look forward to more publications on the achievements of women in WASH to complement existing efforts and create a society where water is prudently used in sustainable ways.

Eng. Sylvain Usher
Former AfWASA Executive Director
Advisor of the AfWASA Executive Board
African Water and Sanitation Association (AfWASA)
Côte d'Ivoire

The Women in Water Project

During the International Water Association (IWA) World Water Congress and Exhibition held in Tokyo in 2018, the IWA Specialist Group on Sustainability in the Water Sector organized a workshop focused on staff training, knowledge management, and organizational culture issues. After the final facilitated discussion session, two shy young women walked up to one of the discussion moderators to express appreciation for the information the workshop had provided. They said that they would like to try out some of the ideas they had heard. But they observed that they wouldn't have the opportunity to apply these in the utility where they worked. They explained that in their country, water utilities were run by men who weren't interested in hearing any new ideas from women; male managers wanted female employees to do as they were told.

This conversation inspired the Management Committee of the Specialist Group to think seriously about the role of women in the water industry. Providing information to water professionals on how we could improve our work in the water industry was worthwhile. However, providing information at conferences would not be enough if structural barriers imbedded in the corporate cultures of water organizations prevented their female employees from having sufficient influence to put good ideas into practice. We wanted to learn and understand more about existing barriers and their consequences, while working to alleviate them through a combination of research and professional development efforts.

We approached the problem from two perspectives. First, barriers that prevent women from bringing all their knowledge, skills, and capabilities to their work are a gender equity issue. Second, in the case of jobs in the water industry, the failure to fully utilize everyone's capacities is also a survival issue. There is a serious gap between the magnitude of the water supply and water quality problems we face and the level of people-power that water organizations are currently bringing to the task. Given the water and climate change challenges of our time, we can't afford to waste anyone's knowledge or skills. Full utilization of everyone's capacities is a goal that needs to be accepted and acted on by the water industry, not only for the sake of women, but in order to fulfil our responsibilities to the communities we serve and the planet we need to protect.

The Specialist Group's efforts to date have included multiple events (three webinars and a workshop), which have been combined with research efforts designed to deepen our understanding of problems and possible solutions. This book reflects our findings to date. The information we have received from speakers and research respondents indicates that the industry is changing. Women in the water industry are participating in a broadening variety of roles, using knowledge and skills developed through education and experience. However, progress is occurring unevenly, at a pace and level of urgency inconsistent with the challenges we must address.

Speakers and research respondents discussed their own experiences and what they had observed in the careers of others. Although the speakers in the webinars and workshop described in this book were female, men from the water industry attended the events, responded to our surveys, and provided useful input to our surveys. We hope the book will serve as a window to provide a more complete view of where we are, as well as a bridge to additional research, development of useful professional development materials, and wider, deeper communication about how to address our shared challenges. In order to make the findings of this project accessible to as many individuals as possible, this book has been published Open Access (available on the internet free of charge).

We believe the insights we have gained through research to date will provide a sound basis for building new tools to support improved collaboration between women and men in the water industry.

The Structure of the Women in Water Project

The work covered in this book includes three webinars and one workshop, with speakers and research respondents from all over the globe. The first workshop was held on Women's Day in 2021, with an international cadre of speakers. The webinar held on Women's Day in 2022 focused on Latin-American countries; presentations made in English were translated into Spanish, and presentations presented in Spanish were translated into English. The workshop was held at the IWA World Water Congress and Exhibition in Copenhagen in 2022, with support from AYSA Water and Sanitation utility in Buenos Aries, Argentina. The webinar held in 2023 was co-sponsored by IWA and the African Water and Sewerage Association (AFWASA); all speakers worked in the water industry in Africa.

The webinars and workshop were designed to provide encouragement and guidance from both experienced female water professionals and Young Water Professionals (YWPs). All webinars were recorded by IWA and are available for viewing at the links shown below.

WEBINAR 1: WOMEN'S DAY 2021

https://iwa-network.org/learn/empowering-women-in-water/

Speakers:

Diana Ulloa
(Ecuador)

Rosie Wheen
(Australia)

Suvritha Ramphal
(South Africa)

Elisabeta Poçi
(Albania)

Norhayati Abdullah
(Malaysia)

Korpo Jensen
(Liberia)

WEBINAR 2: WOMEN'S DAY 2022

https://iwa-network.org/learn/empowering-women-in-water-perspectives-from-latin-america-and-the-caribbean/

Speakers:

Malena Galmarini
(Argentina)

Juanita Ayala
(Sweden)

Geisel Sanchez
(Costa Rica)

Blanca Jiminez Cisneros
(Mexico)

Sharon Archie
(Trinidad and Tobago)

Martha Orta Zambrano
(Ecuador)

IWA WORLD WATER CONGRESS AND EXHIBITION IN COPENHAGEN: 12 SEPTEMBER 2022

Speakers:

Eugenia Ghiotto
(Argentina)

Diane D’Arras
(France)

Titilola Bright-Oridami
(Nigeria)

Louise Dudley
(Australia)

Cheryl Davis
(United States)

WEBINAR 3: WOMEN’S DAY 2023

https://iwa-network.org/learn/women-in-water-africa/

Speakers:

Chataigne Djuma
(Democratic Republic of the Congo)

Maggie Momba
(South Africa)

Leunita Sumba
(Kenya)

Maha Salah Khallaf
(Egypt)

Geraldine Mpouma Logmo
(Cameroon)

The project included research activities such as the following:

- A survey conducted prior to the development of the first webinar
- Surveys conducted with individuals who registered to attend webinars
- Questions asked by attendees during webinars
- Answers provided by panelists either during or following webinars.

Questions asked and demographic data collected in research surveys associated with two of the webinars (one conducted on Women's Day in 2022 and one conducted on Women's Day in 2021) are included at the back of this book: see Appendix.

Through these efforts, the Specialist Group was able to amass a body of knowledge that reflects not only the views of speakers, but also the questions, concerns, and suggestions of both women and men across the globe. This is not a book of statistics; it is a book of voices, including some voices that are rarely heard. In the case of survey responses, we generally know only the information provided by respondents on their gender, country of origin, and (in some cases) whether or not the respondent was a Young Professional. Questions we have asked in surveys have been open-ended, to allow respondents to express the ideas most important to them.

We have organized the data we collected from our varied information sources using the following categories:

- Career Opportunities and Advancement
- Disrespectful Behavior
- Self-Confidence
- Impact of Age on Challenges Experienced
- Work/Life Balance
- Mentoring and Staff Development
- Attributes and Perspectives of Women
- Opportunities
- Next Steps.

We have tried to bring diverse speakers and views into the book, so that readers can benefit from the voices that speak most directly to them.

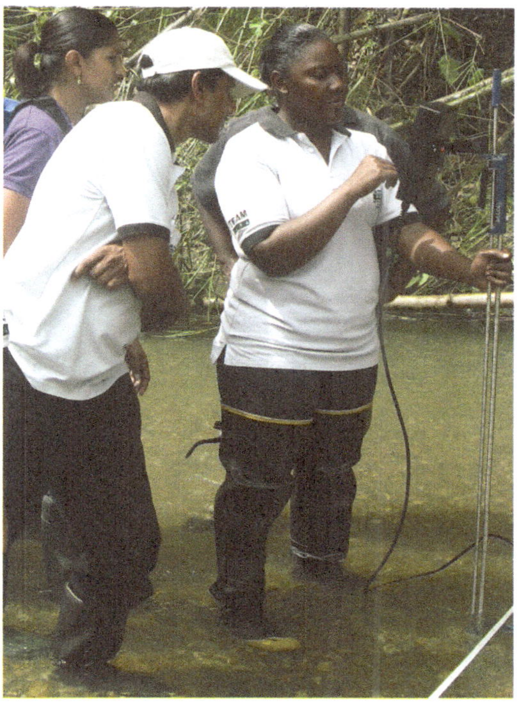

Findings

1. CAREER OPPORTUNITIES AND ADVANCEMENT

In many work environments, expectations about the role and capacities of women affect their ability to participate fully in the planning and implementation of organizational activities. As a result, women are often limited not only in the type and level of jobs they have the opportunity to hold, but also the degree to which they are able to contribute their ideas, knowledge, talents, and energy to the job they have. This is illustrated by some of the responses provided by respondents to the survey question, "What are the gender equality challenges faced by women in the water sector in your region?"

A Young Professional from India commented, "When women today are able to fly high and fight for the country, then working vigorously in the water industry shouldn't be a big deal. The major challenge for women in this field is that they are under-estimated and criticized, rather than their capacity." Another Young Professional from India echoed this view: "I believe that if a person is well-educated and trained in a particular field, there must not be any barriers for that person to work anywhere in the industry, be it a him or her." Another female respondent from India stated that she personally felt that women had the same capabilities as men; however, she reported a personal experience with gender discrimination: "I was not allowed to sit for an interview for a job. After arriving for the interview, I was told that males candidates were preferred because the job involved field work. I was flabbergasted that they didn't call and inform me if this was the preference."

A Young Professional male from Denmark observed the effect of exclusion of women from training opportunities, as well as the impacts of social expectations on the women who did attend: "There is a lack of representation of women in online webinars, seminars, conferences, etc. With regard to participants from the Middle East, women often do not feel confident enough to engage in asking questions or sharing their experiences (which would be valuable to the audience)."

A woman from the United States reported that she had worked for an organization in which the women were often overlooked for senior management in one component of the organization: "We still have men in the industry who are very dismissive and/or patronizing, when the woman is the expert in the area."

Opinions of respondents on the question "What gender equality challenges are faced by women in the water sector?" included the following:

- From a woman in Chile: "Although women are part of the water workforce, they are not properly represented in leadership and decision-making. The water industry in Chile is tremendously masculinized, which affects the quality of decisions made regarding this resource."
- From a female Young Professional in Ecuador: "Living in a rural indigenous community means living in the midst of machismo. Community leaders do not believe or trust in the potential that a women has to solve problems related to water in an engineering way."
- From a female in Iran: "Gender discrimination in obtaining water-related management jobs."
- From a man from the United States: "Getting enough women and minorities interested in the profession and training them. This is compounded by the long-held but wrong belief that it is not work for women, and the social and institutional barriers to females entering the field, starting with education."
- From a female Young Water Professional in Mexico: "There are many challenges to career advancement for women in many sectors in the economy, including the water industry, although gender discrimination is decreasing."
- From a woman in Grenada: "Very few females are involved in the water sector outside of administration at the utility companies. I have to work very hard to earn the respect of males."
- From a woman from Brazil: "Most of the people in high positions in the government of Brazil are men."

- From a woman in Bolivia: “Lack of technical skills/training (especially in the area of civil engineering). Women generally have less field experience than men. There is low confidence in female leadership.”
- From a female from Portugal: “Gender discrimination in obtaining water-related management jobs.”
- From a female Young Water Professional in Guyana: “Males are primarily in charge since women are rarely interested in pursuing this subject because men will always be given the top jobs.”
- From a female in Suriname: “Men like to rule. They do not like it when I have a great idea. They judge it, and if they like it, they make it seem like it was their idea.”

One of the challenges women often face in terms of career advancement is a question about their leadership capability. In response to a survey completed prior to the Specialist Group’s first webinar on Women in Water, a woman from Bolivia commented that staff “lack confidence when a woman is in charge of making decisions”. In response to the same survey, a woman from the Netherlands commented that women face an “extra need to prove yourself and gain acknowledgement”. A Young Professional woman from Chile pointed out that generally field work was done by men, and that her biggest challenge had been “to lead groups of mechanical and electrical installers and expect them to trust my decisions and abide by them”. A survey respondent from Greece noted that there were fewer women in higher ranked jobs, and that there was prejudice (usually from males) who “don’t respond well to women engineers giving instructions”. A consultant from the Philippines reported, “There is low trust in women’s capability and work stability because women are always perceived to be fickle-minded and known to make high-risk decisions.” A male survey respondent from Nepal made the following observation about the leadership of committees of water users and sanitation committees in Nepal: “There is a provision for electing females to key positions in water users and sanitation committees in Nepal, but in some cases the female elected members

Suvritha Rampal, South Africa: "Women are still under-represented in the water and sanitation sector in South Africa. There are regulations, legislation and laws in place to help change those dynamics in the country, especially in private companies, Yes, we are under-represented, and yes, we have created an enabling environment to change this, but not fast enough. There are cultural, social norms that we have to change. How women perceive themselves has to change as well."

are dominated by male elected members, due to lack of confidence in their leadership capacities."

Some respondents commented on water organizations that limited the types of work that were offered to women. A female researcher from Malaysia expressed frustration at limitations on the roles within water utilities that women were expected to perform. "I am a researcher and find it quite challenging to go into the wastewater plant sector. It seems most of the guys are assigned to treatment systems and the ladies are assigned to the office department." She noted that it would be a great opportunity to allow women to explore water treatment as a career by giving them a chance to work in the plants. She thought it would be good for them to learn the reality of the work which they were currently only monitoring from the office using computers. Similarly, a woman from Grenada noted "Very few females are involved in the water sector outside of administration at the utility companies". A survey respondent from the UK said, "Many jobs in water operations are still considered 'male' careers." A Young Professional from Norway also noticed that there weren't many women working in water treatment plants.

Maha Salah Khallaf, a webinar speaker from Egypt, discussed disparities in assignment of leadership roles that couldn't realistically be attributed to differences in the capabilities women and men bring to the industry:
"In the water sector in Egypt, we have about 25 water and wastewater companies. Of 25 companies, how many would you guess have a female chairman? One female chairman out of the 25 companies. I had a chat with one of the decision-makers in the sector and asked, 'Why is there only one woman? Why only one, why not five?' And the answer was, 'They cannot stay late.' I said, 'In medicine they stay late. In many of the other disciplines, they do stay late', so, I assume it's because there is no encouragement. Those who are making the decisions are just not convinced that the women are good enough or equally good for this position. Unfortunately, I have to say that if I look at the organization I work in, an international organization, how many women are leading projects: water projects, wastewater projects, agricultural projects? Not even 25% of the number are female. So, I think that as female leaders in the sector, we should encourage other female leaders and we should also encourage the sector to change."

Water organizations also have a responsibility to pay women fairly for the jobs they currently perform. Unequal pay for equal work was mentioned by many survey respondents. For example, a Young Water Professional woman from Ecuador cited the need for equal pay for women performing operational jobs in the water industry. A respondent from the United Kingdom also reported resistance to obtaining financial compensation that was commensurate with her skills and experience:

"The water industry in the UK is very male dominated. It is really hard to enter as a woman, even with a PhD in engineering from a leading university, and very difficult to be promoted. When I moved to the UK, I had three years of working experience but was paid less than British males with similar experience. After I finished a PhD, companies only offered me entry-level jobs, ignoring my five years of experience and my four years of work on a PhD. An interviewer at a well-known engineering consultancy asked me during an interview what job my husband had and why I needed to work as I had a husband in a good job."

A female survey respondent from Sweden noted that there was a 10% difference between the average engineering salary for a woman and a man.

A female respondent from Nicaragua said, "First, women are the ones who carry the water in buckets. Second, few projects take their participation into account. And for the women who work in the sector, there is wage inequality."

A Young Professional from India observed that women got paid less than men, even in entry-level positions.

The barriers that stand in the way of broad career opportunities and equal pay for women often begin in the home and the education system. The pattern continues from there into the workplaces in the water industry. Equity requires changes in social, educational, and business norms, as well as active efforts by women to make full use of their knowledge and talents.

Geisel Sanchez from Costa Rica offered these recommendations:

"It is important to get involved from the student or entrepreneurship stages in different community initiatives, in non-government organizations, in grassroots organizations, to learn the area of interest of each one. This allows us to have a better vision of the territory and allows us to develop skills that will be important at the time of exercising our professional roles. This allows us to have better tools that lead to empowerment. In addition, it is necessary to continuously train in leadership, politics, communication, and teamwork, whether in national or international spaces.

At the university level, it is necessary to promote the admission of women into careers that are considered masculine, for example, engineering. By working on real opportunities for women to get involved in a wide range of careers we can ensure their participation in the water sector. In addition, it is important to empower women as students, so that participation in the sector does not depend on their leaving the university, but can be included beforehand in research or projects, as well as in training and workshops.

Opportunities in the level of study and job offers vary by country, but if you have an interest in being part of this sector, I can only recommend getting involved. Do not wait for them to call you, look for spaces, apply for places, and participate actively. We require women engineers in sanitation, managers, coordinators, and project leaders, as well as women in management positions in public and regional companies related to water. Participate in youth networks. In Latin America there are quite a few in each country."

Titilola Bright-Oridami, a workshop speaker from Nigeria, also encouraged women in the water industry to take responsibility for continuous learning:

"In the fast-paced world of the water industry this is not just a choice, it's a winning strategy that sets you apart from your peers. Embrace the mantra of 'knowledge is power' as it has been the foundation of my successful journey. Keep yourself informed and up-to-date with the latest trends in the industry through continuous professional development. By investing in your learning, you will witness growth and find yourself leading the charge toward excellence."

While it is true that women can and should actively prepare themselves for important roles in the water industry, it is also true that rapid progress in this area will not occur unless the current leaders of water organizations actively open and clear the way for their progress. Efforts toward change must be top-down as well as bottom-up. Given the higher percentage of men in leadership positions in the water industry, this means that change must be supported by men as well as women.

2. DISRESPECTFUL BEHAVIOR

Many speakers and research respondents reported situations in which they were treated disrespectfully by male bosses or colleagues, or observed female colleagues being treated in an unprofessional way.

Elisabeta Poçi from Albania shared this experience from her early days in the water industry:

"I remember something that will always remain with me. In the early steps of my career, I had prepared for a meeting. I was in charge of preparing all the reports and materials that were to be discussed and I remember I had prepared everything to perfection, every detail was checked. And it was a big table in our office meeting room, which was full of men. I'm tiny as a person and my voice is small. Please take my size into consideration, because it makes a difference. And there was a moment, when the discussion was heated. I hadn't spoken a word. But I felt I had to clarify something. So, I spoke up in that room, although I was intimidated. I could not even hear myself. That's how loud the discussion was. And there was this man sitting right next to me, who turned to me and said: "You do not have the right to speak." And I was shocked. I was very young and that was the last thing that I needed to hear in those early days. I was still trying to prove myself to myself, not just to other people, and that was really shocking. I keep the memory as a souvenir, but I would like to tell young women who might face similar situations, to not be intimidated by such behavior. You earn your place around the table, and your right to speak, by the value you contribute to it, and nobody can take that away from you."

Diana Ulloa, a speaker from Ecuador, described a meeting in which a boss attempted to 'put her in her place' by assigning a role that he considered appropriate for her gender, instead of a role consistent with her expertise and job responsibility:

"When I was a Technical Adviser to the highest level of Water Secretariat in Ecuador, there was an important meeting where almost everyone attending was a man. I was in my office preparing technical information and was nervous before this important meeting. Have you ever noticed a woman being nervous and reviewing a lot of information before a meeting? It was not a common meeting: there were some Ministers, managers, and the Executive Director. But not one of them had studied anything related to water. I was from the unit that had studied water engineering and water management. Although I had a Master's degree and I was the Technical Adviser, my boss started the meeting by saying to me, 'Darling, you will take notes while we share some ideas in this meeting.' For me these words meant, 'You don't have anything to say in this meeting, no matter who you are, what you've studied, what you know. You are a woman, and you will take notes. We don't respect you.'

I was afraid, but I said, 'I am not going to take notes. You have your assistant to do that.' In some moments, you have the chance to choose if you want to do what people want you to do or respect yourself and say no. It is difficult to say no, especially for women, and especially in an environment where everyone thinks you have to do what your boss says, but this is a step toward having self-esteem and becoming a leader."

Maggie Momba, South Africa: "In everything we do, there are challenges. When I started a post-graduate program, it was a big issue. The environment I was working in was not conducive for women. The atmosphere was characterized by jealousy. The men didn't like a woman to even turn on the process controller. But we were born to leave a legacy. Seek collaboration, but don't let anyone bully you."

Although these comments reflect experiences from the early days of the careers of three webinar speakers, the questions and responses of many individuals who participated in the research activities associated with the project showed that the issue of disrespectful behavior toward women in the workforce place is a current reality for many women in the water industry.

A Young Professional woman from South Africa reported that group meetings were sometimes still used as an opportunity to illustrate the lower status of women: "I have been sidelined, been asked to stick to my lane, or asked to take minutes in a meeting where I am the Chair."

A Young Professional woman from Brazil reported "prejudice, sexual harassment, and devaluation" at work.

Geraldine Mpouma Logmo, Cameroon: "We are in a company led by men. This is the reason we set up a women's network. We needed to be heard. When we go to a meeting, women will not talk. Men only will talk, talk, talk. And not because women don't have anything to say, it's because they're afraid. They are not even willing to take the floor. And some men are not even giving the floor. Our director was even cutting, taking the floor from me. I said, 'come on, let me finish, bear with me. Let me finish speaking. And then if you don't agree, we can see'."

Sometimes reservations about a woman's capacity are subtly expressed. A female consultant from Latvia said, "Sometimes it is a silent requirement to prove your knowledge, experience and skills."

A Young Water Professional in Norway who worked as a consultant reported that she "was not taken seriously. I have experienced this from both my manager and my clients (for example, not being copied on correspondence, or a client speaking directly to my manager even though I'm the project lead. I have experienced that my work is double-checked or that my ideas don't go through as easily as the ideas of male colleagues."

> *A female water professional from Chile felt that her work experience included "prejudice on the part of managers, peers, subordinates and clients about technical capabilities of hard engineering in design, in the field, in the execution of projects, physically managing resources, correcting errors, or proposing improvements to the work of a male colleague. In work meetings, competence must be demonstrated to the point of impressing others in order to gain the credibility which a male colleague has earned by the simple fact of being seated at the table. Seriousness and capacity must be demonstrated permanently, even when already occupying managerial positions earned by merit, since it is also necessary to break down the rumor that the position was not obtained by professional capacity. It must constantly be made clear that we are not trying to seduce anyone."*

The above response was submitted as a response to a survey question, and the specifics of the personal interactions that generated it are not known. However, it is notable and unfortunate that a female worker would feel permanently constricted and defensive about her behavior in the presence of male colleagues after being informed that rumors were circulating to the effect that she had obtained her position for reasons other than professional merit. It is difficult for women to develop collaborative relationships with male colleagues in such a difficult work environment.

Disrespectful behavior toward women was also noted in a survey response from a male attendee of one of the webinars, a man who worked for a government organization in the United States: "Sometimes men (especially in leadership positions) don't value, listen to, or leave space for women to contribute equally. An example of this is when someone talks over a woman or only listens to what a man is saying."

Bullying of women was reported as a common occurrence in some portions of the water sector.

A female professional from the United Kingdom made this observation: "The main barriers I see are still around the 'Old Boys' network and the prevalence of bullying in some consultancy and construction sectors. The Water Companies themselves I think have removed most of this, but it does still exist in some of the less progressive companies. I am hopeful that with the emphasis on stamping out all types of inequality this will help, but I think too little time is spent in actively managing and mentoring people in the consultancy and construction industries to sort out problems and break down barriers. Too many people turn a blind eye."

Titilola Bright-Oridami, a webinar speaker from Nigeria, reported her own experience with disrespectful behavior:

"I have been in such situations, and my response has consistently been to handle them with the utmost respect. Whether it involves respectfully declining, avoiding involvement, or requesting a department transfer, I firmly believe in taking a stand against disrespectful behavior. Moreover, it is essential never to engage in such conduct ourselves."

A female Young Professional from Canada who responded to a pre-conference survey said, "In the water industry, I have worked in environments that were hostile toward women. I think the lack of women leaders made it difficult to speak up and to be understood when I did speak up."

It is essential for all employees to have an opportunity to contribute their best work.

Martha Orta Zambrano, a speaker from Ecuador, addressed this issue:
"Among the main difficulties faced by women is to be heard at the working tables. However, with character and conviction, change can be achieved. To do this, it is important to always be trained and to contribute with innovative ideas that demonstrate the commitment and passion that you have in each action. I recommend having a positive attitude towards adversities, and looking for solutions to the problems that arise. The role of women in water and sanitation projects is key - having a vision, being versatile and innovative, taking care of every detail of projects from their planning and execution to implementation with the community."

3. SELF-CONFIDENCE

In the context of this study, self-confidence is defined as having an accurate sense of what you know and still need to learn, combined with confidence in your ability to learn what you need to know in order to make a valuable contribution. Mature self-confidence makes it possible to request input and assistance from others when help is needed, as well as offer support and assistance to others. This level of confidence also makes it possible to offer ideas in more difficult circumstances, such as assignment to a junior position in the organizational hierarchy, or in meetings where managers and supervisors resist input from women.

Self-confidence can also help also women avoid defensiveness and anxiety in situations where they want to perform well, but still have a lot to learn. Lack of self-confidence can be reflected in unwillingness to ask questions because of the perception that lack of information will be perceived as weakness. In reality, the water industry is so complex that everyone in the industry has a lot to learn. Fear of asking questions can be a significant detriment to individual learning, as well as an impediment to providing effective supervision and management. Water professionals should be asking questions and learning continuously, and it is valuable for supervisors and managers to model this behavior.

Self-confidence is also essential in moving past mistakes, whether real or perceived. In the water industry, there are many opportunities for things to go wrong. This reality can be particularly difficult for women when their competency is already in question. A survey respondent from Brazil observed, “We still suffer from some types of prejudice. Recently there was a major accident that broke sewer pipes and social media blamed the women involved in the project. The great repercussion showed there is still much prejudice against women in sanitation.”

Prejudice can predispose peers, managers, or the public to blame a woman for problems, but the reality is that everyone makes mistakes. One aspect of self-confidence is to have the capacity to acknowledge mistakes and address consequences calmly, honestly, and energetically. In either women or men, insecurity can lead to cover-ups or blaming others. Developing a level of calm that does not rely on perfect performance (or a pretense of perfect performance) is important preparation for career advancement. Higher levels of responsibility involve a higher level of visibility, and a level of authority that makes it possible to make mistakes with serious consequences. Leadership roles also require managers to take responsibility for mistakes made by others in their chain of command. Since many female respondents to the research survey associated with this project expressed a desire to become leaders in the water industry, this aspect of strong leadership – being able to admit to mistakes and deal responsibly with the mistakes of others – should probably be included in training materials developed to help both men and women develop the skills needed for effective supervision and management.

The amount of self-confidence that women bring into the workplace often reflects the attitudes of the families in which they were raised. Geisel Sanchez, a webinar speaker from Costa Rica, believed that she had benefited from the good examples of her grandmother and her mother, noting that in the house where she was raised, "a career had no gender". She knew that she could study environmental engineering. But the university presented challenges: "This taught me that sometimes I should raise my voice, talk in a strong way. Empowering yourself and professional networking can help you deal with uncomfortable situations."

Whereas women who feel insecure about their role in the workplace sometimes respond by being defensive, women who are self-confident find it easier to maintain a professional level of communication, even in difficult circumstances. Learning how to deal successfully with conflict is essential for career success, since disagreements and power struggles are inevitable components of the world of work.

Blanca Jiménez Cisneros from Mexico, whose experience includes work in academia, water utilities, non-profit organizations, and government diplomacy, reflected this in her response at a webinar to a question from an individual who attended. The question of the workshop participant was:

"I understand what you are saying about changing the way we support and resist each other–you can get more with honey than with gall. But doesn't continuing to behave as a pretty and smiling woman contribute to the stereotypes with which

we have been pigeonholed? We continue to be branded as hysterical when we express ourselves assertively."
Dr Jiminez's response was "I agree about being pretty and sweet being a woman's stereotype, but being firm, kind, and tough with a smile is not the same. Frequently, because of our weariness related to the jokes, challenges, and abuse that we face, we react in the wrong way. The balance is difficult, but we can achieve it with practice. There are no general recipes because it depends on the context and the culture to which one belongs."

Women in a male-dominated industry or work group sometimes feel uncomfortable. A female Young Professional from the UK remarked, "Often being in the minority on teams and in discussions causes intimidation and a greater pressure on women to prove themselves. Feeling like they are somewhat 'other' can cause insecurity." When commitment to the work itself is a shared priority, it is possible for men and women to work together in a situation that feels like 'us', rather than 'us' versus 'them'. However, in some situations, women feel that male peers are more committed to proving them wrong than having the work go right.

Self-confidence doesn't involve over-estimating your own knowledge, skill, or accomplishments. It does involve being willing to come forward with your questions and beliefs. For women taught as children to defer to others, this can be a challenge.

Rosie Wheen, a speaker from Australia, provided guidance on how to develop self-confidence:
"For me this has been one of the journeys that I continue to go on in terms of leadership. I think self-reflection is so important. I've really come to understand for myself what some of the self-talk is that holds me back from speaking out when I want to, what keeps me from challenging issues with confidence and really holding my power. When I was a young girl, what was rewarded was to be quiet, to be polite, not to question. Whereas now, as a leader, that's exactly what I want to do. Of course, I want to be kind and respect others. But I also want to challenge how things are and to change them. And I would really notice in myself sometimes that self-talk of 'Uh-oh, don't say that. People might not like you.' And now, recognizing something that's been socialized into me frees me from that story in many ways. Sometimes I use little tricks to get myself feeling confident, things like listening to music lift my energy. I love using a power pose to give myself strength. Have a go! Try standing like Wonder Woman

for a minute and feel that power rushing through you. Doing some of those things to trick our body into feeling more confident has an amazing impact. I often do that before I'm about to do a big talk."

Geraldine Mpouma Logmo from Cameroon pointed out that confidence could be undermined by professional harassment from male colleagues:

"Thank God I did not face sexual harassment. But professional harassment, I have faced. As a woman you always face such things. Lack of patience, what men tolerate and don't tolerate with us. They think we should do more because we are women. Maybe at the end of the day, I think we should work on the confidence of women. I don't know why we women are afraid of making mistakes. This is part of us." Her advice to women is "We need to really focus on work and skills. Don't pressure yourself too much. Give yourself time to learn."

Leunita Sumba from Kenya also reinforced the importance of self-confidence. "You need to believe in yourselves, your capabilities. Identify your niche. I have identified water pollution and lack of safe sanitation, and I'm doing something about it. As we are all aware, women and girls are the most affected when water and sanitation are lacking. I am pushing for changes in policies, practices, attitudes, and mindsets in my activities. Being a CEO, you really need to be confident, but it doesn't just come. You must start building it even when you are still in a junior position. I had to reinvent myself. That meant doing a lot of speaking up and being visible. So, the confidence has come with time and I'm still not there, but it has improved over time." Noting that her own career had included "phases of growth interrupted with phases of challenges," she acknowledged that setbacks had the potential to undermine self-confidence. However, she noted that she had been able to reach out beyond the job setting to obtain a broader perspective. "When my job as a CEO came to a not-so-good ending, the book by Dr Wale Akinyemi, *Help, My Chocolate is Melting!*, came to my rescue, with a story told to him by his mother. When a chicken lays an egg and you take the egg, does it cry? No, it lays another egg. So instead of crying when I lost my job, I asked myself, 'Do you still have eggs in you? Yes.' I moved on, and I have kept laying more eggs."

4. IMPACT OF AGE ON CHALLENGES EXPERIENCED

Some participants in this project reported discrimination based on age and inexperience, paired with an unwillingness by male colleagues to contribute to the opportunity for women to learn.

Chataigne Djuma, Democratic Republic of the Congo: "At the beginning of my career, I would be told, 'You have no experience, you are too young for this' or 'You are a woman, what can you do when you are working in the field?' but I believe now that people are beginning to understand that gender equity is a non-negotiable human right."

A female Young Professional from Sweden who worked for a company that supplied technology to the water industry reported significant difficulties:

> *"As a woman I have experienced both sexual harassment and discrimination based on my gender. In my ten years in the industry times have changed, and many managers are now willing to promote women because of a shortage of female leaders in the water industry. However, being a young woman, I experience difficulties with subordinates every day, mostly from older men who experience my supervision as painful and problematic. Many men are not ready to listen to strong female voices yet, unfortunately."*

Sharon Archie, a speaker from Trinidad and Tobago, experienced age discrimination as a more experienced worker at the utility where she worked, when a manager was committed to allocating training opportunities to younger female employees.

"As I was moving up on the hierarchical ladder of the company, at a certain point I noticed that my manager gave preference for training to younger female colleagues. I said to the manager that he should share the training for all ages within the company. He actually told me to get out of his office. He continued to provide the training to younger colleagues. I did not receive any more training in those last years, until he left the company a year ago."

5. WORK-LIFE BALANCE

Many speakers and respondents, including men, noted the work-life balance issues confronted by women in the water industry. Work-life balance issues included childcare, household responsibilities, care for physically infirm family members, and the possible negative impacts of professional achievements and income on a woman's ability to attract a husband and have children. These issues were discussed not only in terms of the judgments of others, but also women's judgment of themselves and their female colleagues. This was also an area where many respondents saw a need for changes in organizational policies and benefits to make it feasible for women to balance their families and their jobs successfully.

A woman from Peru observed in her response to a survey question about challenges faced by women, "There is still the thought that mothers cannot exercise the role of leaders because of their work in the home." Currently, in most if not all cultures, women are hard-pressed to handle the challenges of leadership in the workplace when they are also taking primary responsibility for work in the home. That is particularly the case when domestic responsibilities include childcare and the necessity to perform household tasks without the benefit of modern conveniences. While any water professional can be challenged by work-life balance issues, it is common for women to be expected to bear more than half of the responsibility for raising children and the health needs of other family members.

Norhayati Abdullah, a speaker from Malaysia, discussed challenges faced by women in the water industry who have family responsibilities:

"We need to eliminate the stigma that has hindered potential growth among women professionals in the sector. Women with family responsibilities should be seen as equally capable of taking on challenging tasks. Opportunities should be made available within the water sector for women to be much more involved, not only at the working level but also in decision-making positions. The shining nature of a capable woman at work is perceived as deteriorating the moment she progresses in her personal life.

There is a stigma that we can never progress professionally simply because we have kids, or we are pregnant, or we are tiny. I'm tiny, and wherever you go to the table dominated by men you feel out of place, you just can't bring yourself to the front despite having talents that you can share with the rest of the group. This stigma can come from others or from your own mindset. One factor is having male colleagues who are not supporting or sponsoring the progression of female colleagues in their careers. But women might also think themselves that they are incapable. The moment we have children we might think we are not capable to do work because we have small babies. Or we might say to ourselves, 'I'm not capable of doing that because I'm breastfeeding'. Or 'I'm not capable of doing it because I have small kids and I have to do school runs.' This is basically all in the mindset, which is not allowing us to see ourselves as progressing in expanding our professional careers."

Maha Salah Khallaf, a speaker from Egypt, made these observations and suggestions based on her own experience:

"In the area where I live, in the Arab-African region, if you are a young woman and looking to pursue your education in engineering, the first

thing you are told is this: this will hamper your opportunities to have a married life. If you are enthusiastic enough to go to a university and study engineering as a post-graduate abroad, you will be told, 'You know, if you do your post-graduate abroad, this will limit your chances in finding a husband.' Because in this culture if you have a Master's degree, you're expected to marry someone with a Master's degree. If you have a Master's degree from abroad, your husband or partner should have a degree from abroad. Not to mention a PhD degree. So, when I went to pursue my Master's degree in England, I was told, 'Be careful, it will limit your marriage chances.' This is alarming for many ladies. And if you start work, the question is always, can your salary in this culture be higher than your husband's salary? And if you work for an international association, sometimes you must travel and leave your kids behind, leave them to someone else to take care of them. So, there are always these questions about limitations. Once you survive these choices and you decide, yes, I would like to pursue all the choices that are open for me and use them, you must live with your choices, which is very interesting.

In order to have all these roles (working woman, mother, partner), what should I advise? My advice for you is to have mercy on yourself. Things are not meant to be perfect, although women are generally perfectionists. And also, we feel guilty that if my kid is sick it is because I am working. No, your kid is sick because all kids get sick, not because you are a working woman. If the house is not perfectly clean, it's because I'm working. No, it's because some days it can't be perfectly clean. So have mercy with yourself and other female colleagues.

My second piece of advice is to prioritize and select what cannot be replaced later in time: a child, or a career opportunity. If it cannot be replaced, pursue it, and then prioritize other things on the list.

My third piece of advice is to delegate. You can always have your stamp on the tasks that you delegate. The opposite of being a perfectionist is being practical; try to be practical.

Last but not least, when you go on your journey, take the whole you with you. We consist of different parts that make us happy. Being a working woman makes me very happy. The part where I do my hobby of designing jewelry makes me happy. When you take over a part of your role, like the mother role, unplug the working woman. If you are in your office, unplug the mother role and leave it in your house. Try to do the role you are currently pursuing the best you can. In your whole life, do not leave a part behind. That will make you happy."

Titilola Bright-Oridami, a speaker from Nigeria, also acknowledged both the difficulty and the importance of finding a work-life balance that is consistent with personal happiness:

> *"As a woman, finding your unique equilibrium among work, family, and personal life becomes a transformative journey. It's about embracing the art of balancing responsibilities while safeguarding your well-being. Prioritizing self-care and ensuring your overall health remain paramount as you navigate the beautiful tapestry of life. Remember, work-life balance is a journey, not a destination. It takes a lot of time and effort to find what works best for you. But it is worth it to create a life that is full and satisfying in all areas."*

In the survey responses that were received in the course of the Women in Water Project, many respondents expressed concern about how women could handle family responsibilities in the context of current expectations about (1) work hours and (2) time off allowed in connection with family responsibilities. For example, one Young Professional from South Africa provided the following response to a survey question about solutions to improve gender diversity in the sector:

> *"I think it's firstly important to make sure that women feel safe and accepted. That the environment is conducive to the presence of women who work hard at work and then go home and work harder for their families. The working hours used do not have a woman, a caregiver, mother, and nurturer in mind. The tools used in environment health and safety have not been adapted to suit women."*

There are some work settings where personnel policies have been modified to address this challenge. **Malena Galmarini**, a utility manager from Argentina, discussed gender policies that had been implemented in her utility for parental leave, improving working conditions for men as well as women:

"A more equal world is extremely important. That's why we got to work right away on a cross-cutting gender policy at AYSA. We started with the prevention of acts of violence – institutional, labor and personal violence. But we didn't stop there. We advanced in the training and awareness of all the personnel of all the companies that we work with and of the community cooperatives that carry out many of the works in the most populous neighborhoods. We aimed to reach gender parity in positions, in functions, but also income. We are moving forward with licenses for co-responsibility. This means inviting the workers, the men of the company, to be co-responsible in the upbringing of their children. We did not only extend maternity leave, but also extended paternity leave from 10 to 30 days. We also expanded caregiver leave from 24 to 32 hours to promote co-responsibility in children's care. The latter makes it possible for fathers to accompany the mother at school events, parents' meetings, the doctor's office, a soccer game, or a play. This incentivizes women and men to be part of it all in an equal manner."

Louise Dudley, another speaker with utility management experience (in this case, in Australia) also endorsed helping employees meet their family responsibilities as well as their responsibilities at work: "It's important to create a work environment that is sensitive to work-life demands – one that encourages people to perform work they value, while offering the freedom to enjoy the things that matter most in their lives."

6. MENTORING AND STAFF DEVELOPMENT

Mentoring was often mentioned in feedback, whether in the context of yearning for additional support and guidance, an acknowledgement of mentoring provided by family members or colleagues, reporting on mentoring provided to others, or encouragement for development of support groups. Chataigne Djuma, from the Democratic Republic of the Congo remarked, “We don't have a lot of mentors who are trying to help us understand what is going on.” Similarly, a woman from Canada who contributed to a survey said, “In the water industry I have worked in environments that were hostile toward women. I think the lack of women leaders made it difficult to speak up and also to be understood when I did speak up. I also think the mentorship I received was much more limited for me compared to my male counterparts.”

A female Young Woman Professional from Belize observed that female environmental engineers were not considered for project and management posts. The experienced engineers had monopolized the projects and didn't invite the participation of young women engineers. Since academic learning needs to be supplemented by on-the-job experience, the willingness of experienced staff to support the learning of beginners can be critical to their success.

Some women in the water industry feel a responsibility to try to meet this need. Diana Ulloa from Ecuador said:

> *“I try to find opportunities to work with women who have less experience in the water sector because it is my chance to create opportunities. It's not that I don't respect men, but I want to help create more women leaders in the water sector. It's difficult because you have to invest a lot of time and face challenges, but it is a good experience because you create a network.”*

Suvritha Ramphal from South Africa said "Observe and understand each other's personalities and characters. Don't take communication at face value–allow what has been said to permeate your mind. Look at different angles of what has been communicated to you. Communicate respectfully. Give each other encouragement.

One of our superpowers is that we are emotional. We just need to understand how to manage it. It helped me to find an older individual within the sector to talk to. It helps to have someone you can trust, who can guide you and give you advice on how to handle certain situations."

A woman from Mali who provided research input prior to a webinar also noted the support that women could give each other:

"There are many challenges women are facing in the sector. At the professional level, women have less chances to reach professional jobs. At the educational level, women must reinforce their capacities. At the networking level, women must unite and make a strong network to empower each other."

A water professional from the United Kingdon mentioned the benefit to women of finding mentors and champions to support them in their careers, and the benefit of "seeing others like them in senior roles, to encourage them that promotion is possible". She noted that there were not enough mentors to guide younger workers and provide a 'sounding board'. Similarly, a woman from the

Juanita Ayala, a speaker living in Sweden, provided numerous suggestions for achieving positive mentoring relationships:
"Sometimes the mentoring relationship is forged naturally, for instance, by showing curiosity and interest in the knowledge and career path of a professional close to you, like your boss, colleague, or professor. In these situations, build a mentorship relationship by showing interest and proactivity. If you don't know the person you want to have as a mentor, reach out. Politely but casually present yourself and ask what you want to know. Don't make it extremely formal or ask directly for mentorship. Instead, ask for information you really want to know. Remember that this person probably has a very tight agenda. How would you catch his or her attention and make the best use of their time? Then build the relationship from there.
Use social media, Twitter, and LinkedIn. If someone doesn't answer, can't mentor you (for many reasons - time, interest, or even because you haven't been clear enough about what you want or expect), don't be discouraged.

You will get many NOs in your career, and that's normal. That means you are out there looking for opportunities, trying and growing. Remember to be patient and persistent.
Women can feel vulnerable. It is important to be able to ask for help/support/mentoring. This does not mean you are a failure or weak. In those moments of support great ideas can be developed. Mentoring is an empowering way to support other women to progress in the sector and be an agent of change."
In response to a question from a webinar participant on 'What are we going to do to empower all women?', she responded, "Start by empowering yourself. Be in control of your emotional intelligence and lead by example. There will be hard times where you will be discouraged, not motivated, or even demotivated. Remember why you started this path and how far you have come. Building a career is a marathon, not a sprint! Then you and your example will not only inspire but also open new spaces for other women. Share your experiences. Show interest and support for other women."

United States mentioned that in her organization, there were no females in leadership positions, "making mentoring difficult."

Sharon Archie from Trinidad and Tobago said: "We as women often tend to compete against each other. I often tell women we are not here to compete but to complete each other. It appears that there is limited space for women at the top, so that competition begins. But we have to remember that the world is vast and there is room for every skill and competency. We need to encourage team-building exercises and promote gender equality in the workplace and within the community."

Malena Galmarini from Argentina said: "The best way to work as a team is with a feminist focus, breaking the idea of competition and replacing it with a collective construction. The first rebellious act that women can achieve is to trust in other women."

Geraldine Mpouma Logmo from Cameroon encouraged women to be part of networks in terms of mentoring and coaching. "We need role models, we need mentors. And the best mentors I have had were men. I had some mentors that assisted me. So we do have champions." Her observation that some men were supportive was reinforced by a comment on women's capacities made by a male from Mexico, in response to a survey about gender diversity: "They're just as prepared and smart."

A woman from Colombia who completed a survey connected with the webinar geared to a Spanish-speaking audience also had deep appreciation for the support she had received from a male mentor: "I was lucky enough to do my PhD in a lab where my advisor preferred to work with women. I was lucky enough to work on multiple and cutting-edge projects, which, regardless of being a woman, allowed me to develop as an engineer and researcher. It was a single yes from my adviser that opened the door to a whole horizon of opportunities."

While discussions of mentoring often focus on fellow employees in current work settings, it is also important to be aware of, appreciate and seek out broader sources of support, whether from our past, our current lives, or resources we can use to help us develop.

Leunita Sumba noted that by treating their sons and daughters equally, her parents had been her first mentors:

"My story begins with my family. I have three brothers and two sisters. There was no discrimination between the girls and the boys. I say this

because it was not the norm then and it's still not the norm in some African families. My mother advised me to do all of the studying I wanted to do. She was my number one mentor. My mother was my number one female support." At later stages of her life, she felt she needed to be mentored by individuals she had never met. *"A book by Robin Sharma called* The Leader Who Had No Title *taught me to lead even if I did not have a position of influence. Another by Simon Sinek,* Leaders Eat Last *has taught me to sacrifice what is mine and to save what is ours. That means mentoring – at an earlier time mentoring my students, and now mentoring younger women and clearing the path for them on water resources, protection, and conservation."*

From this broader view of mentoring, webinars, workshops, and publications intended to help women bring their best to the water industry can also be defined as a form of mentoring that can be provided even at a distance. However, it is important not to overlook that staff development is a basic responsibility of organizations in the water industry. Industry peers and outside organizations can play a role, but it is also important for utilities and other water organizations to support the training and career growth of their employees.

7. ATTRIBUTES AND PERSPECTIVES OF WOMEN

One interesting aspect of the research findings associated with this project was the degree to which respondents reported a belief that women could bring particular value to the water industry because of their greater awareness of the impact of water problems on women and children. For example, Chataigne Djuma from the Democratic Republic of the Congo said, "Women and youth need to be at the helm of the water and sanitation sector to accelerate achievement of Sustainable Development Goals."

Korpo Jensen, a webinar speaker from Liberia, described her awareness of the impacts of water problems on women and children as a strong motivator to work in the water industry despite the many challenges she experienced: "The story of my beloved country inspires me. With a population of 5.2 million inhabitants and the many challenges faced by my nation, I am eager for a changed future. Of said number, less than 10% have access to tap water for public or private consumption. Of the immeasurable challenges, however, there is a vocal need for institutional capacity. There are few water professionals playing strong roles in a technically male-dominated sector, hence, there is a need for strong female participation in every sphere of the sector. Most decisions made are politically driven. My entry into the water sector and knowledge gaining, has been an opportunity to help save the lives of women and children, by providing services in slum and rural communities. I want to encourage female workers not to give up on the task at hand."

Many female survey respondents noted the disproportionate impact of water quality problems on women and children. A woman from Uruguay noted:

"In rural areas in general it is women who provide water to the family, but when there is an opportunity to define policies, they are not consulted. Water is considered a man's thing. Girls also drop out of school due to lack of water and hygiene at school. In rural areas, being isolated and less globalized, changes are much slower. Likewise, the centrality of capitals and cities makes any problem in rural areas invisible."

Similarly, a woman from Brazil noted the lack of women's participation in decision-making within water organizations, but also commented on the impact of water impact issues on the daily lives of women: "For the minorities, access to water supply is still chaotic, and women are over-burdened to solve the day-to-day issues."

A female Young Water Professional from Columbia observed:

"Access to good quality water in the country is a problem that does not discriminate based on gender. However, the role attributed to women in society makes them more vulnerable to this scarcity, mainly in the care of the home and care of others."

A female respondent from Indonesia noted that poor water quality required women to buy clean water for drinking and cooking.

A female Young Professional from Bangladesh stated:

"Women have less opportunity to earn income. The responsibility of carrying water from a distance (usually 30 to 50 minutes' distance) usually falls on women and children. Schooling of children (especially girl children) is hampered due to water carrying. Personal hygiene (especially of adolescent girls) is not maintained properly. The safety and security of girls while carrying water from a distance is also at risk."

Even women in the water industry are not exempt from water-related problems. A Young Professional from India noted that female employees in the development sector are also affected:

"When women go for long hours of field work in the development sector, the absence of clean and safe washroom facilities prevents them from relieving themselves adequately. Females always have to go with male colleagues for field work due to the unsafe conditions at the site."

The disadvantages experienced by women sometimes escape their notice because they are so familiar.

Eugenia Ghiotto, a speaker from Argentina, raised the issue of the 'extra cost' paid by women in their lives and in their daily activities. She noted that it was important to recognize and talk about those extra costs because "what is not said does not exist."

In her presentation at a workshop held in connection with the 2022 IWA World Water Congress and Exhibition in Copenhagen, she observed that her aim was to "make visible these inequities that we as women suffer and generally we can't see—the reality we live every day without noticing the costs." Her suggestion for attendees at the workshop was, "Close your eyes, look back, and think about the inequalities you have lived or which you witnessed this week. I would invite you to make those situations visible in your closest environment, at home, at school, so that any woman would not have to pay an extra cost for choosing her own destiny freely."

Despite the challenges they face as individuals, Giesel Sanchez from Costa Rica has observed impressive efforts by female utility staff and volunteers to improve water for everyone:

"I work with women community water managers in Costa Rica and the associativity and solidarity with which they carry out their work

helps other women to carry out valuable actions in the communities and allows them to solve problems in a truthful way, since as partners they analyze and solve situations together. This is the case, for example, of the women in the Union of Communal Aqueducts of Grecia, called UNAGUAS.

In addition, around the country there are many women who lead community management processes through associativity and strengthen integrated water management without any economic remuneration, they are simply women who support and give their time so that hundreds of communities in the country have access to drinking water and the country has high access rates of around 95% throughout the country."

In addition to sensitivity to the water challenges of women and children, some speakers and respondents believed that the positive attribute that women often brought to their work was a particular sensitivity to sustainability. A female survey respondent from India offered the following observation:

"I think that we should have an inclusive education system which should focus on environment sensitivity and sustainability at all levels of education. Creating an awareness from an early stage is the key to achieve success in the long run. And motivation, encouragement, and support to girls and women intending to work in this area will create a better impact, as every woman has a natural bent towards caring for nature."

Martha Orta Zambrano from Ecuador said: "Today's women, with their dedication and passion, have demonstrated that their knowledge, leadership, and resilience, are a key part of the change that humanity is experiencing, within a framework of sustainability and innovation."

8. OPPORTUNITIES

Some speakers and respondents provided examples and suggestions for improving both the work life of women and families and working relationships between male and female colleagues. These are being highlighted for further consideration as progress continues toward a fuller utilization of the knowledge and skills women can bring to the water industry. One suggestion made by a survey respondent from Mali was for women to work together and support each other:

"There are many challenges women are facing in the sector. At the professional level women have fewer chances to reach responsible jobs. At the education level women have to reinforce their capacities. At the networking level women must unite and make a strong network to empower each other."

As observed by Geraldine Mpouma Logmo from Cameroon, solidarity among women does not preclude collaboration with men. Instead, she recommended communicating with men about challenges faced by their female colleagues:

"Women don't have the same weight, are not addressed in the same manner, are not treated in the same manner. We have to face this situation and work on this in our communication. We have a series of activities that we use to build teams. For instance, on Women's Day in Cameroon, we organize a week of sensitization. We invite men so that they can hear our issues. We have some webinars, some workshops, and discussion. Thanks to technology, we can have webinars even though we are not together. We communicate, we stand together, we don't exclude anybody. Sometimes women have activities together and exclude men. We need to include them. We need to find champions around us. We can send some messages towards those who have discriminating or disrespectful behavior. Communication is at the heart. We need to talk to the engineers and to all categories, to the top management, all levels."

Louise Dudley from Australia also encourages a broader view:

"I think it's important for the industry to look at the issue of equality more holistically. Rather than addressing it solely as one of gender balance, the focus needs to be on fostering a culture of diversity and inclusion, where difference is respected and embraced, and where an individual's contribution, knowledge, skillset and experience are valued and recognised. As a role model for women, I think the best thing I can do is to encourage a culture where difference and diversity are valued, where everyone feels able to bring their best self into the workspace every day."

Next Steps

The data gathered by the IWA Specialist Group on Sustainability in the Water Industry documents many ways in which current beliefs and practices relating to gender limit the contributions of women:

- Insufficient support for girls to pursue education that will prepare them for work in the industry (including in some cases insufficient provision for hygiene in schools)
- Insufficient encouragement for women to learn technical skills or perform technical, operational, or management jobs
- Limited mentoring for women
- Failure to provide equal pay for women and men in comparable jobs
- Inadequate support for employees responsible for care of family members (such as children, elders, and family members experiencing health problems)
- Disrespectful behavior
- Inadequate involvement in decision-making
- Failure to provide rewards and opportunities for advancement that are commensurate with capabilities.

Diane D'Arras, Past President of the International Water Association has noted that women's challenges can be both external and internal. "Although men's mentalities are in many cases the reason women are held back, women can also create their own limitations. And the chains can be invisible, coming from education and habits."

While these recurring patterns can be correctly defined as gender equity issues, they may also reflect a failure on the part of the water industry to grasp fully the severity of the water supply, water quality and climate challenges we face. Everyone's best contributions will be needed to protect water, our communities and the planet. Limitations we set on ourselves and others based on stereotypes related to gender and sexual orientation reduce our ability to address our basic survival needs. As an industry, we have both the opportunity and the responsibility to acknowledge the consequences of historical practices, and promote an approach that builds, uses, and appreciates everyone's talents. Diane D'Arras has acknowledged both the difficulty and the importance of the task, but sees progress: "It takes several generations in social rights between the first break-out and the full reality. Working together to develop good practices in the water industry is necessary; it will be a long, long road. But I am confident that women are mobilizing now and that women will get there!"

Given current inequities in the water industry, considerable effort and investment will be required. The Specialist Group on Sustainability in the Water Sector will conduct research to obtain suggestions and the perspectives of both men and women on the modifications needed to mental models, behavior patterns and organizational policies (for example, in relation to pay levels, openness to the participation of women in all job categories, and paid leave policies that help both male and female employees meet family obligations).

Although research often focuses on data and statistics, much human behavior is governed by ideas, concepts, and internal stories. Where behavior reflects cultural patterns that are embedded rather than examined, personal stories can be as illuminating as statistics. We encourage all those with a commitment to gender equity to maximize the skills of every member of the water industry to conduct both quantitative and qualitative research to further define barriers and identify possible solutions.

The IWA Specialist Group on Sustainability in the Water Sector is now moving beyond research into action. Our objective is to create professional development materials that will help men and women in the water industry work together more collaboratively towards shared goals. Mutual support and respect will increase everyone's opportunity to make a contribution. Like water, the talents of our workforce are too precious to waste.

Appendix

Questions Asked in Surveys and Demographic Data Collected in Connection with Women in Water Webinars Provided by the International Water Association in 2021 and 2022

Empowering Women in Water
Women's Day 2021

(1) Country where you work
(2) Age
(3) Gender
(4) Type of Organisation
(5) What challenges and problems have you experienced or observed in terms of the opportunities provided to women to effectively utilize all of their knowledge, skills and talents in their work in the water industry? (Please share specifically personal experience)
(6) What issue relating to the role of women in the water industry would you most like to hear discussed?

Responses

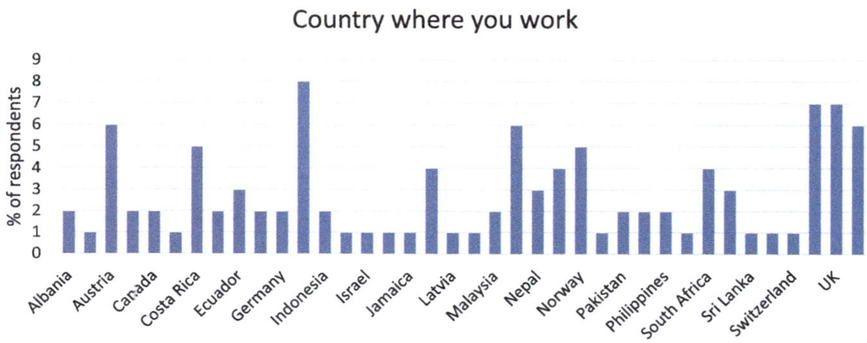
Country where you work
% of respondents
0
1
2
3
4
5
6
7
8
9
Albania
Austria
Canada
Costa Rica
Ecuador
Germany
Indonesia
Israel
Jamaica
Latvia
Malaysia
Nepal
Norway
Pakistan
Philippines
South Africa
Sri Lanka
Switzerland
UK

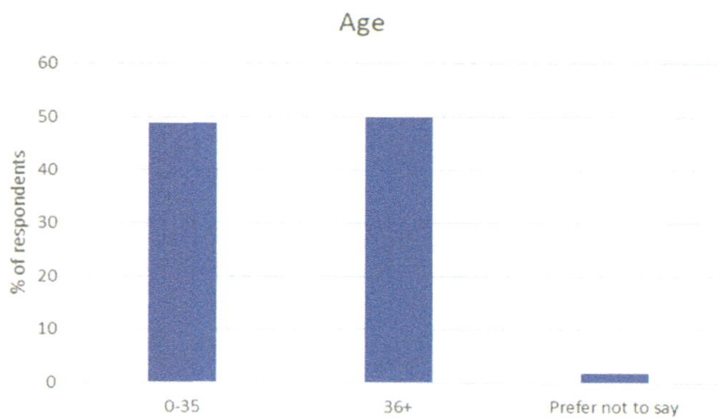
Age
% of respondents
0
10
20
30
40
50
60
0-35
36+
Prefer not to say

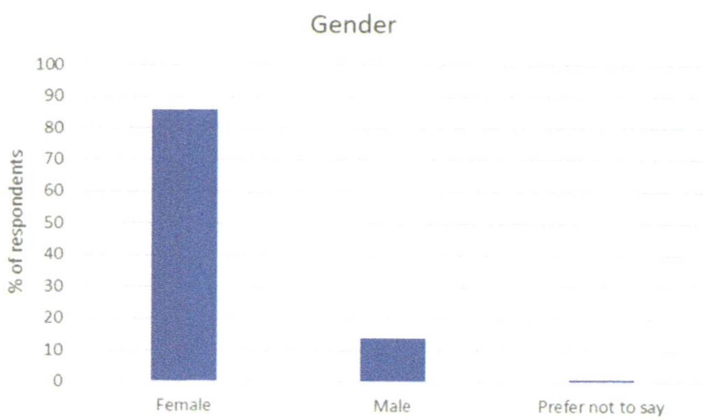
Gender
% of respondents
0
10
20
30
40
50
60
70
80
90
100
Female
Male
Prefer not to say

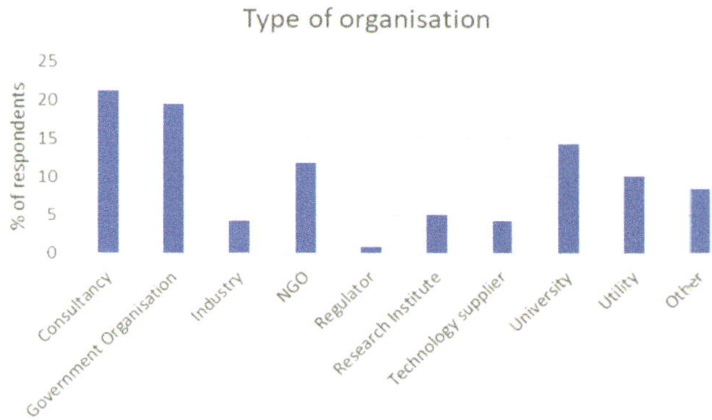

Empowering Women in Water – Perspectives from Latin America and the Caribbean

Women's Day 2022

(1) Email
(2) First name
(3) Last name
(4) Age
(5) Where are you from (country of origin)?
(6) Where are you from (region)?
(7) Where do you live (country)?
(8) Gender
(9) Type of Organization
(10) What are the gender equality challenges faced by women in the water sector in your region?
(11) From your own experience in the water sector, have you faced barriers as a woman?
(12) What do you see as the local solutions to improve gender diversity in the sector?
(13) What are some of the opportunities provided to women to effectively utilize all their knowledge, skills, and talents in their work in the water sector? Provide personal experiences if possible.
(14) What issue relating to the role of women in the water sector in your region would you most like to hear discussed at the webinar?
(15) In your opinion, why do you think it's important to achieve a better balance in the water industry?
(16) If available, please share any relevant references or links related to women in water.

Responses

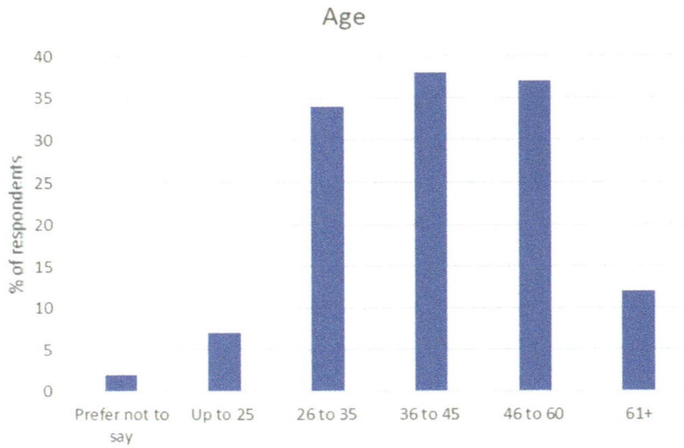
Age
% of respondents
40
35
30
25
20
15
10
5
0
Prefer not to say
Up to 25
26 to 35
36 to 45
46 to 60
61+

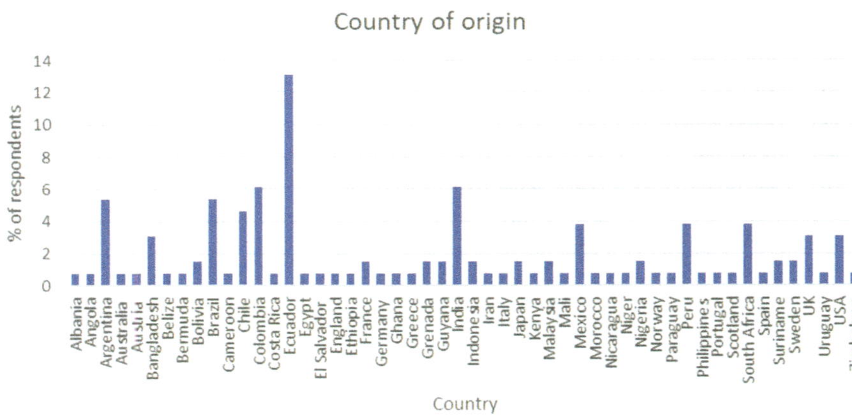
Country of origin
% of respondents
14
12
10
8
6
4
2
0
Albania
Angola
Argentina
Australia
Austria
Bangladesh
Belize
Bermuda
Bolivia
Brazil
Cameroon
Chile
Colombia
Costa Rica
Ecuador
Egypt
El Salvador
England
Ethiopia
France
Germany
Ghana
Greece
Grenada
Guyana
India
Indonesia
Iran
Italy
Japan
Kenya
Malaysia
Mali
Mexico
Morocco
Nicaragua
Niger
Nigeria
Norway
Paraguay
Peru
Philippines
Portugal
Scotland
South Africa
Spain
Suriname
Sweden
UK
Uruguay
USA
Zimbabwe
Country

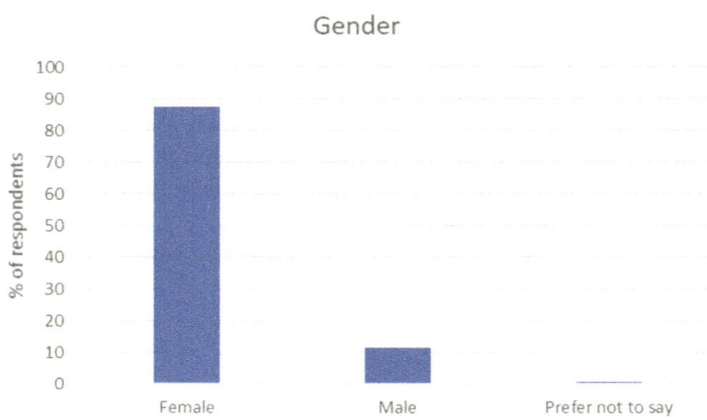
Gender
% of respondents
100
90
80
70
60
50
40
30
20
10
0
Female
Male
Prefer not to say

Where do you live?
% of respondents
12
10
8
6
4
2
0
Albania
Angola
Argentina
Australia
Austria
Mali
Bangladesh
Belize
Bermuda
Bolivia
Brazil
Cameroon
Canada
Chile
Colombia
Costa Rica
Ecuador
Egypt
El Salvador
England
Spain
Ethiopia
Germany
Greece
Grenada
Guyana
India
Indonesia
Iran
Italy
Japan
Kenya
Malaysia
Mexico
Morocco
The...
Nicaragua
Nigeria
Norway
Paraguay
Peru
Philippines
Portugal
Puerto Rico
Scotland
Somalia
South Africa
Suriname
Sweden
UK
USA
Uruguay
Country

Type of organisation
% of respondents
30
25
20
15
10
5
0
Consultancy
Government Organisation
Industry
NGO
Research Institute
Technology supplier
University
Utility

The Authors

Cheryl Davis, Chair of the International Water Association Specialist Group on Sustainability in the Water Sector, served in management positions in the San Francisco Public Utilities Commission for over 35 years. She has also authored publications on topics including water operations, sustainable industrial water use, the water workforce, and ethical issues in the water industry.

Andrée De Cock is an active member of the Management Committee of the Specialist Group on Sustainability in the Water Sector and has been involved in the Woman in Water project since 2021. Andrée co-founded the Young Water Professionals IWA Chapter in Ecuador, and is a post-doctoral researcher at the Aquatic Ecology Research Group of Ghent University in Belgium.

Arlinda Ibrahimllari is a member of the Management Committee of the International Water Association's Specialist Group on Sustainability in the Water Sector, as well as an International Water Association Fellow. She has worked in the water and sanitation industry both as an employee (with a management role in operations) and as a professional consultant. Her focus is on best workforce management practices, as well as empowering women in the water sector and increasing opportunities for Young Water Professionals.

Liudmyla Odud has been a member of the Management Committee of the International Water Association's Specialist Group on Sustainability in the Water Sector since 2019. She has over ten years of professional experience in the water industry, which includes work in the areas of water supply, sanitation, water and wastewater treatment, sewerage, and the environment. Within the Specialist Group, Liudmyla has played a key role in organizing events relating to Sustainable Development Goals, Women in Water, and the contributions that the indigenous perspective on water can provide to the water industry.

Siyka Radilova, a member of the Management Committee of the International Water Association's Specialist Group on Sustainability in the Water Sector, has worked in a variety of sectors, types of institutions, and countries. Siyka is a social anthropologist with extensive experience in Diversity, Equity and Inclusion, with a special focus on gender equality. She is particularly committed to the importance of empowering women in low- and middle-income countries, and within the water and sanitation sector.

IWA Publishing's authorised EU representative for General Product Safety Regulations is Diane D'Arras, 15 rue Duret, 75116 Paris, France, e-mail: safety@iwap.co.uk.

Printed and bound by CPI Group (UK) Ltd, Croydon, CR0 4YY
20/07/2026
02171584-0001